Kiosk

Volume II

Chris Roberts

Introduction

At first, telephone lines and handsets were only installed in the wealthiest of households. Then, they became more widespread and ubiquitous in homes across the population. Already, the decline of the public phone kiosk has begun.

For many years, the kiosk found a new purpose in life: staying connected wherever you were - all you needed was some coins and a number to call. For decades, this purpose was valued by society.

Modernisation was undertaken by British Telecom in the late 1980's - introducing newer kiosks like the KX100 (depicted on the cover) and it's siblings - a modern take on the classic kiosk design, built for robustness against vandalism, and economy of construction.

But technology had progressed. Radio telephony, once the refrain of the rich and frighteningly important, became commonplace and unremarkable. Every person now has the power to connect to almost anyone on the planet, from their pocket.

The kiosks entered another point of their life cycle - in which they are used by vandals to vent their frustration, activists to further their cause with posters and stickers, sex workers to promote their wares, drug dealers to make clandestine deals without traceability.

This second volume of ***Kiosk*** records the decay - the various abuses and indignities which telephone kiosks have undergone as their function is subverted yet further from the one for which they were intended, and ultimately their demise - but then again, perhaps, their rebirth?

Decay

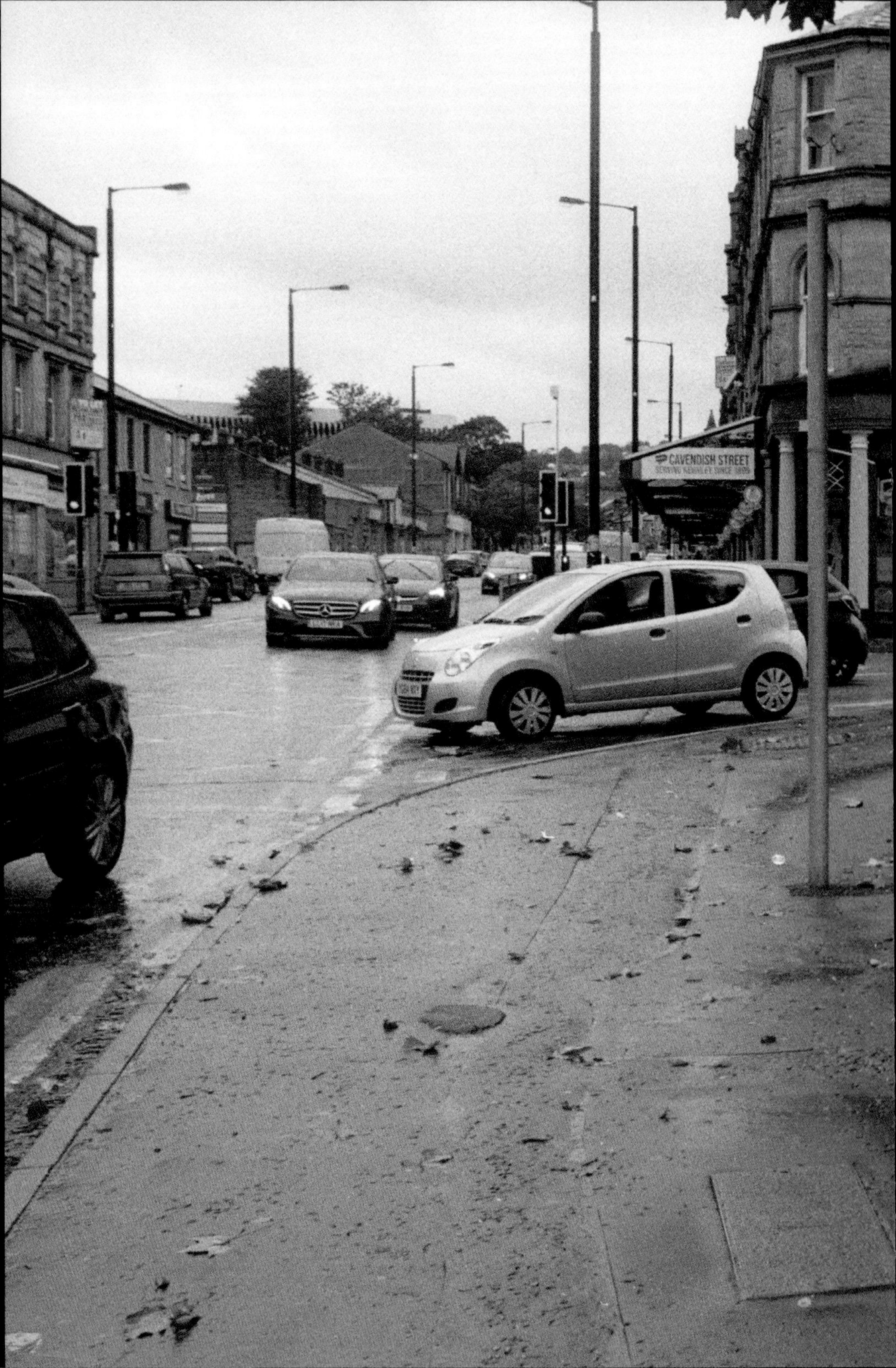

Corner of Cavendish Street, Keighly, Yorkshire

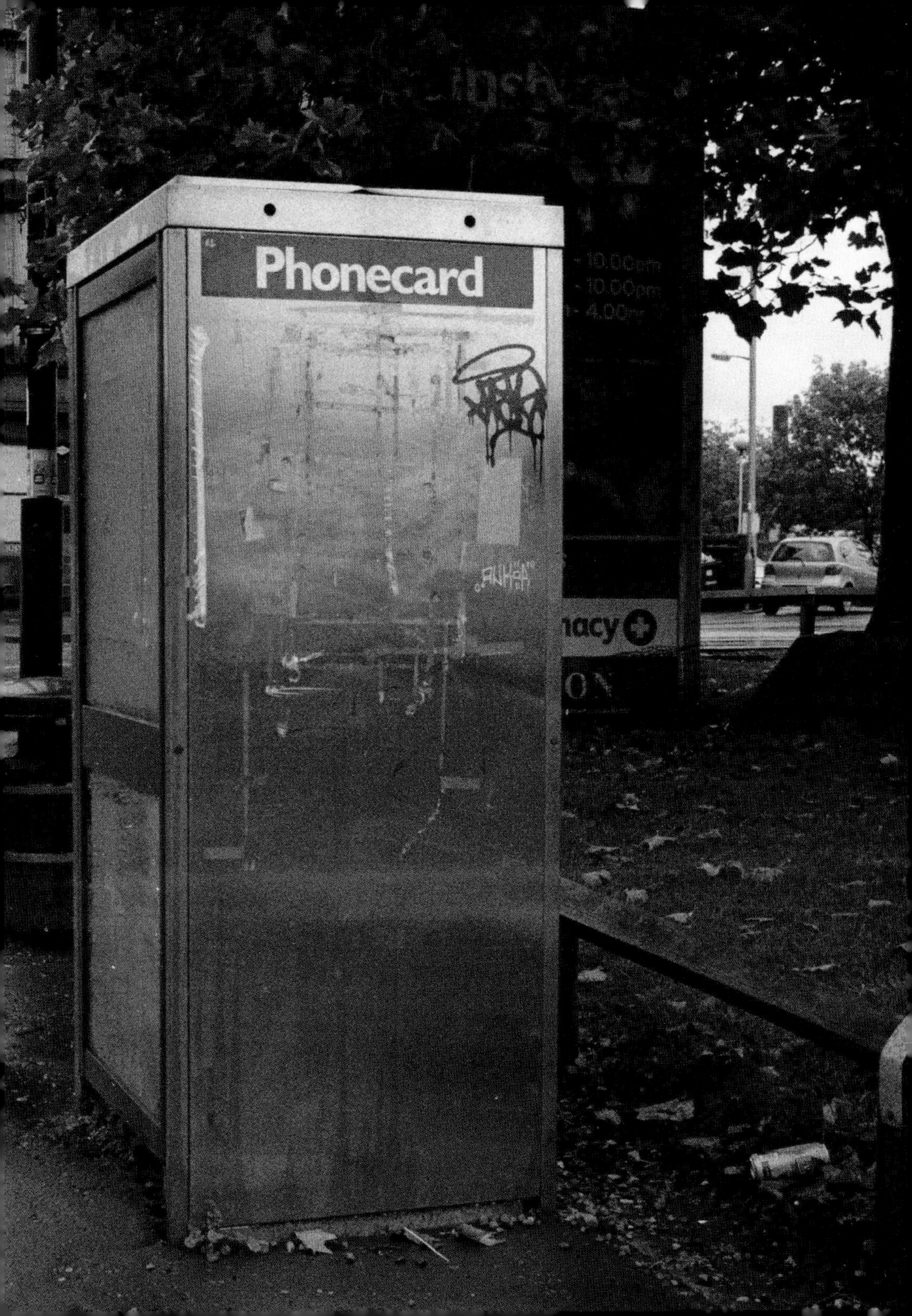
Phonecard

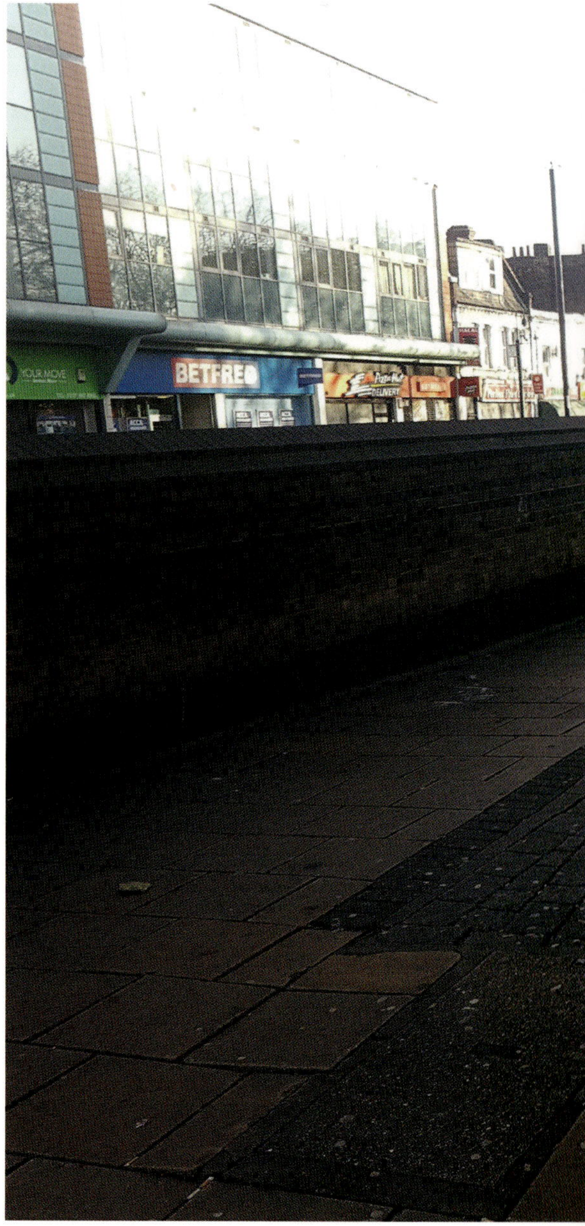
YOUR MOVE
DELIVERY

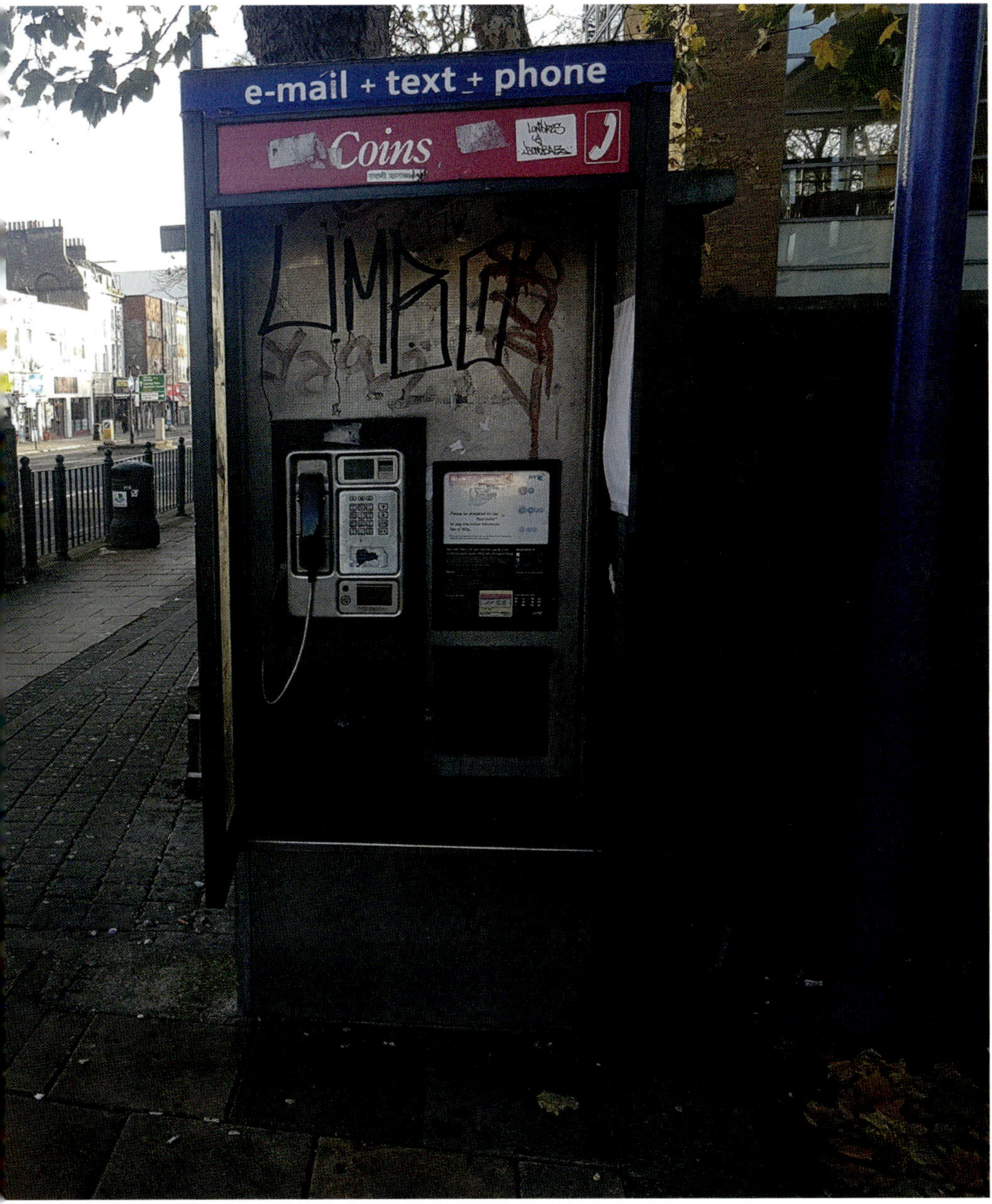
e-mail + text + phone
Coins

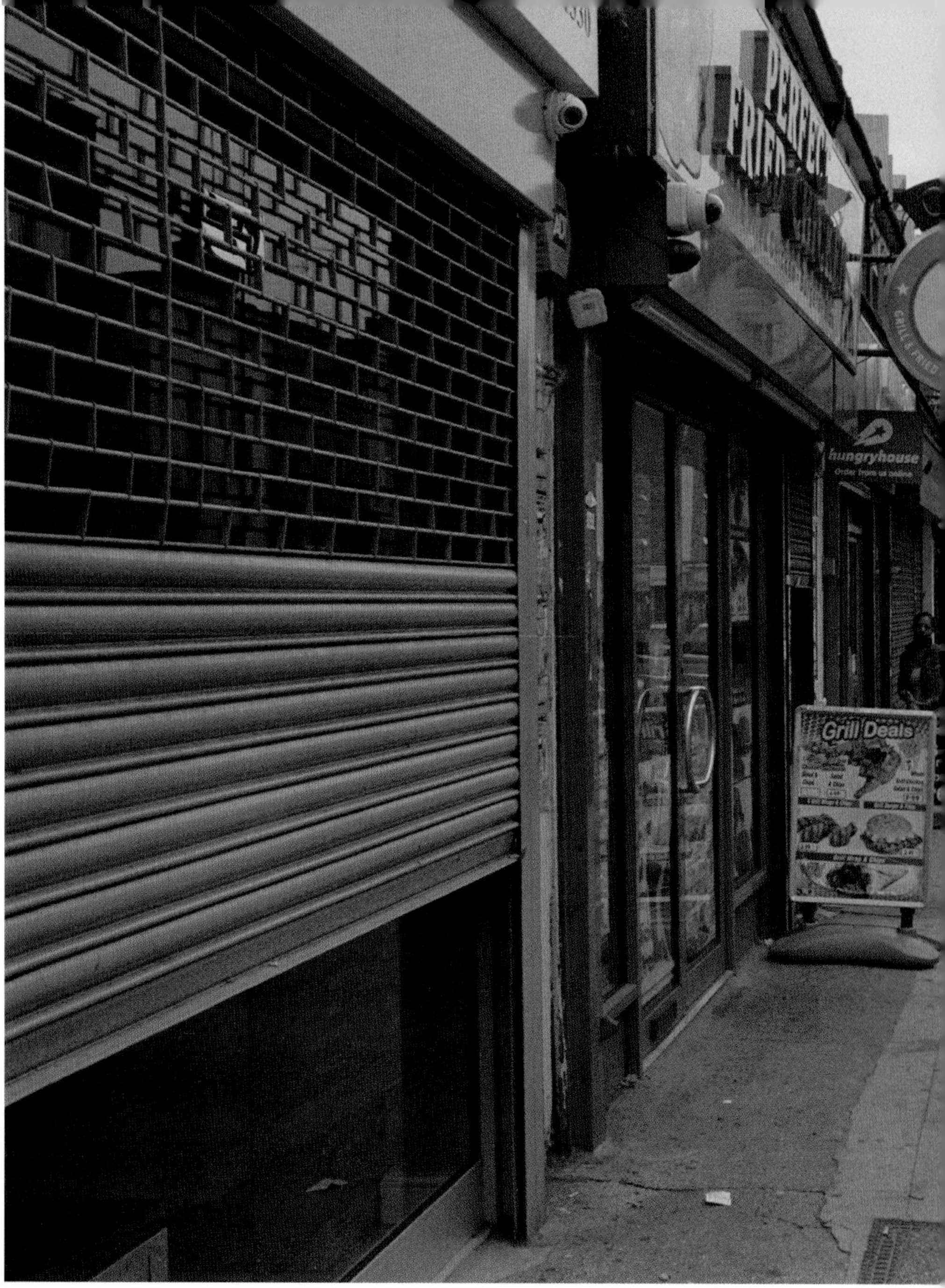

Loampit Vale, Lewisham, London

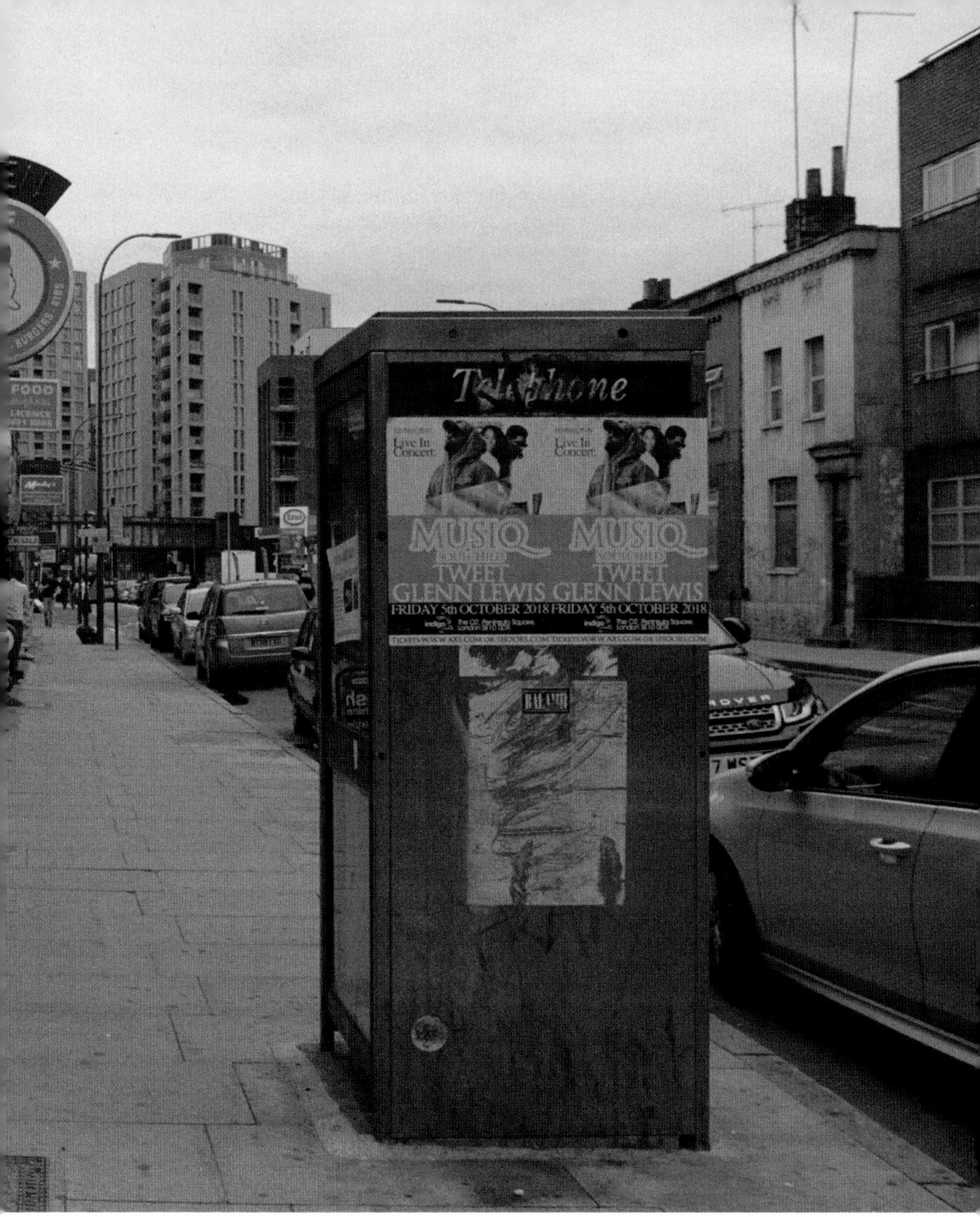
Telephone
Live In Concert
MUSIQ
TWEET
GLENN LEWIS
FRIDAY 5th OCTOBER 2018
Live In Concert
MUSIQ
TWEET
GLENN LEWIS
FRIDAY 5th OCTOBER 2018

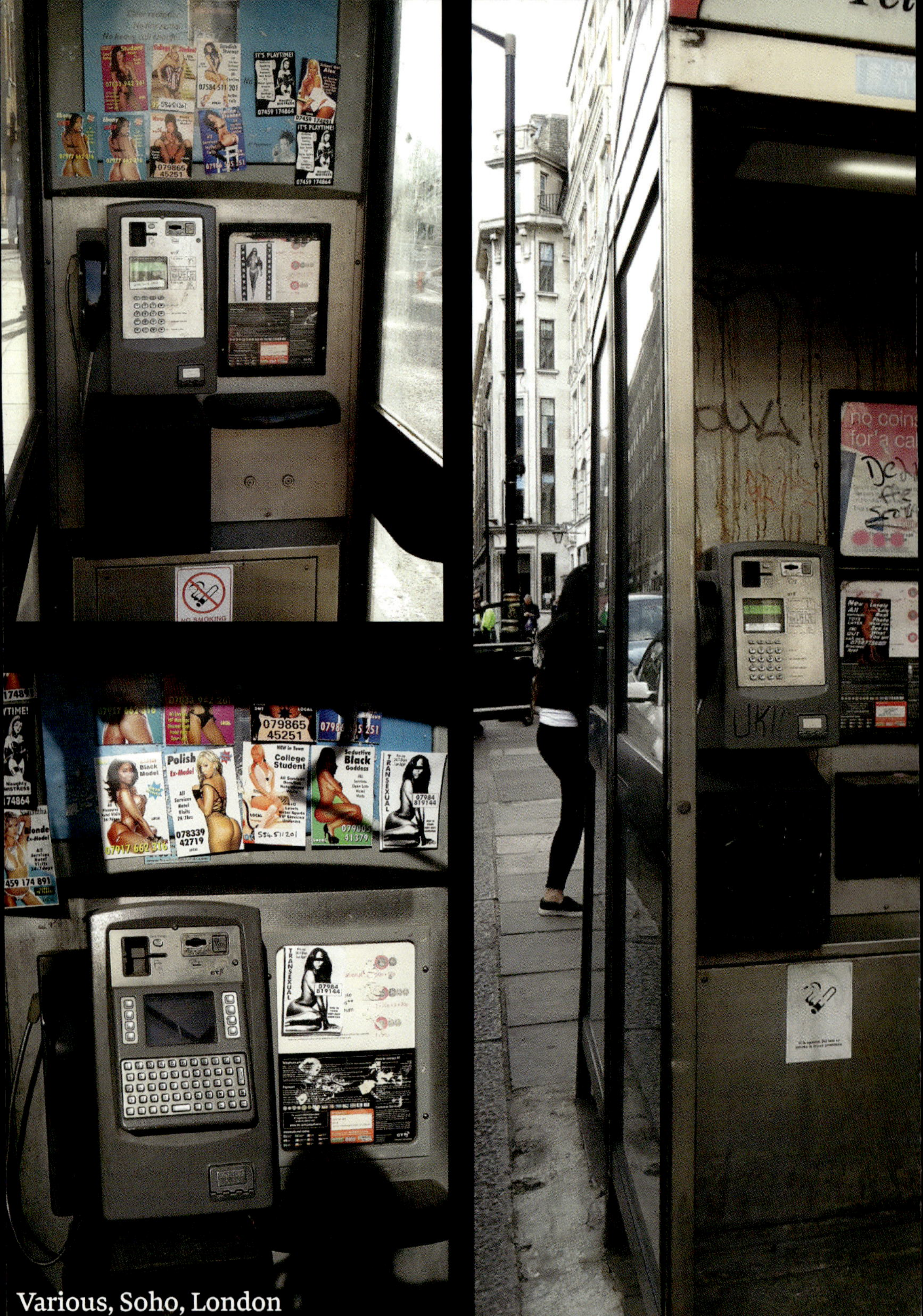

Various, Soho, London

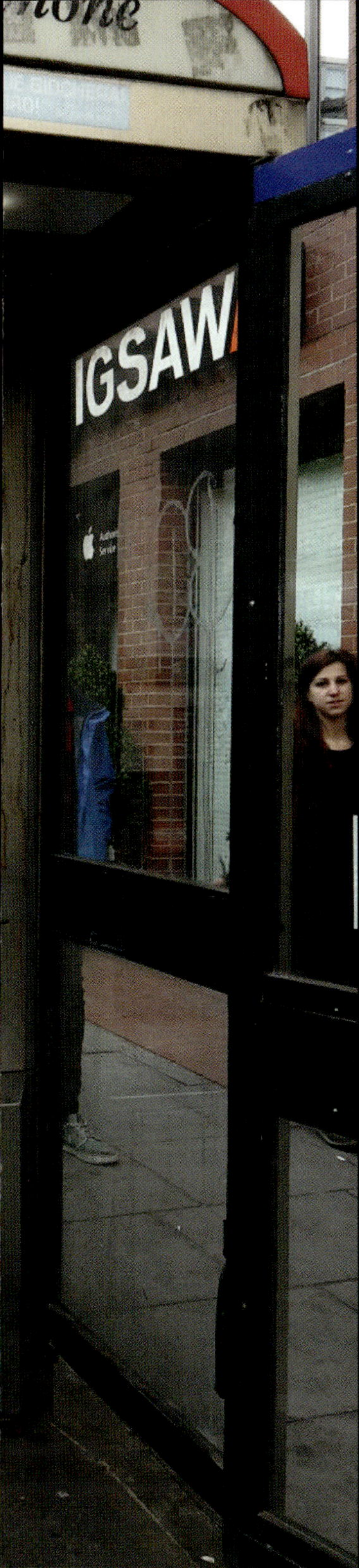
IGSAW

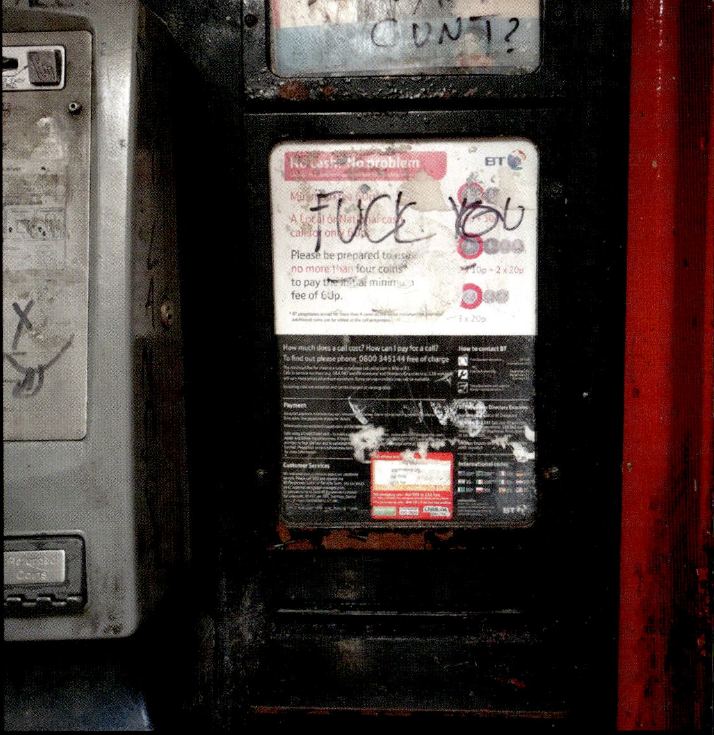
CUNT?
BT
FUCK YOU
Please be prepared to use
no more than four coins*
fee of 60p.
To find out please phone 0800 345144 free of charge
Returned
Coins

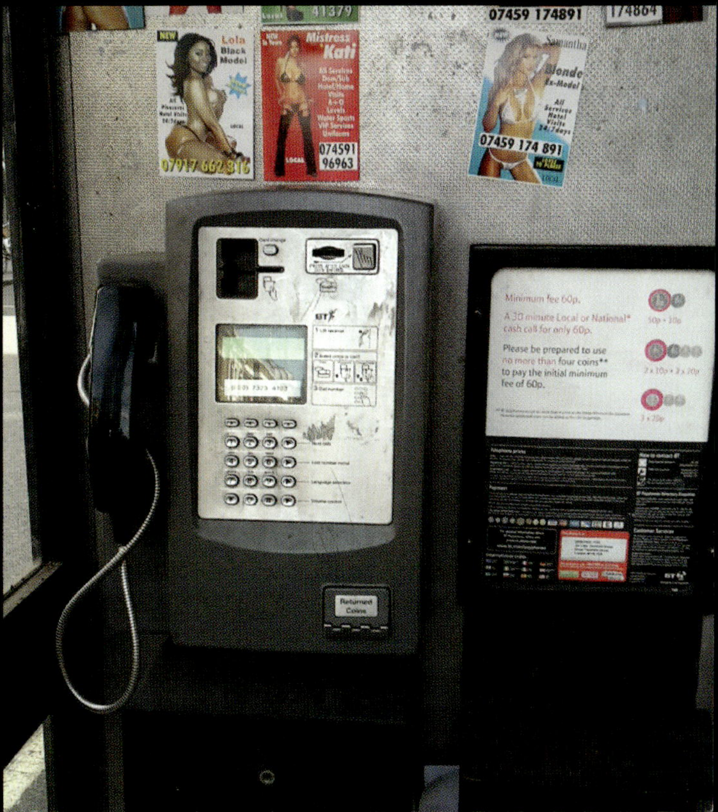
07459 174891
Lola
Black
Model
07917 662 816
Mistress
Kati
074591
96963
07459 174 891
Minimum fee 60p.
A 30 minute Local or National*
call for only 60p.
Please be prepared to use
no more than four coins**
to pay the initial minimum
fee of 60p.
Returned
Coins

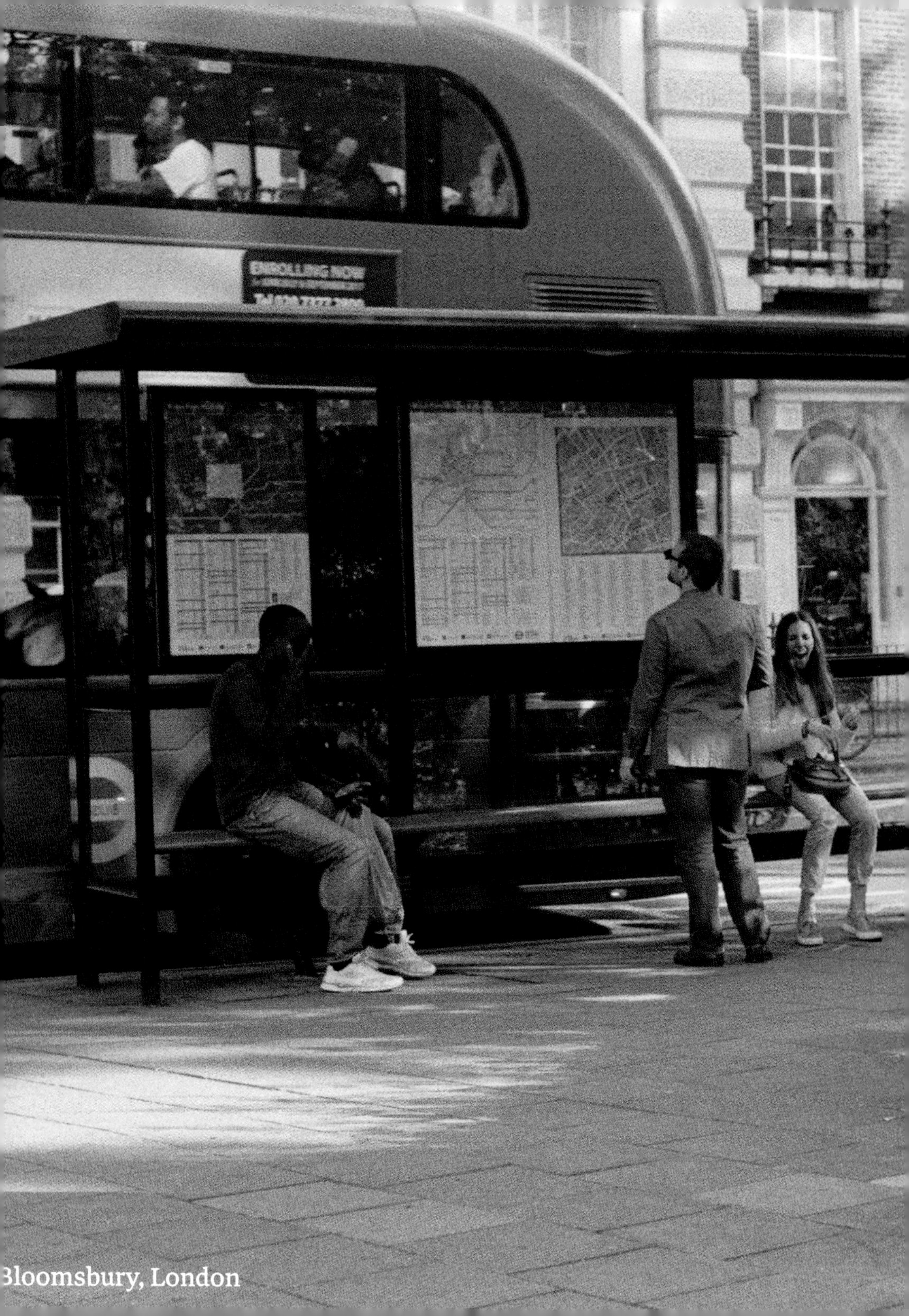

Bloomsbury, London

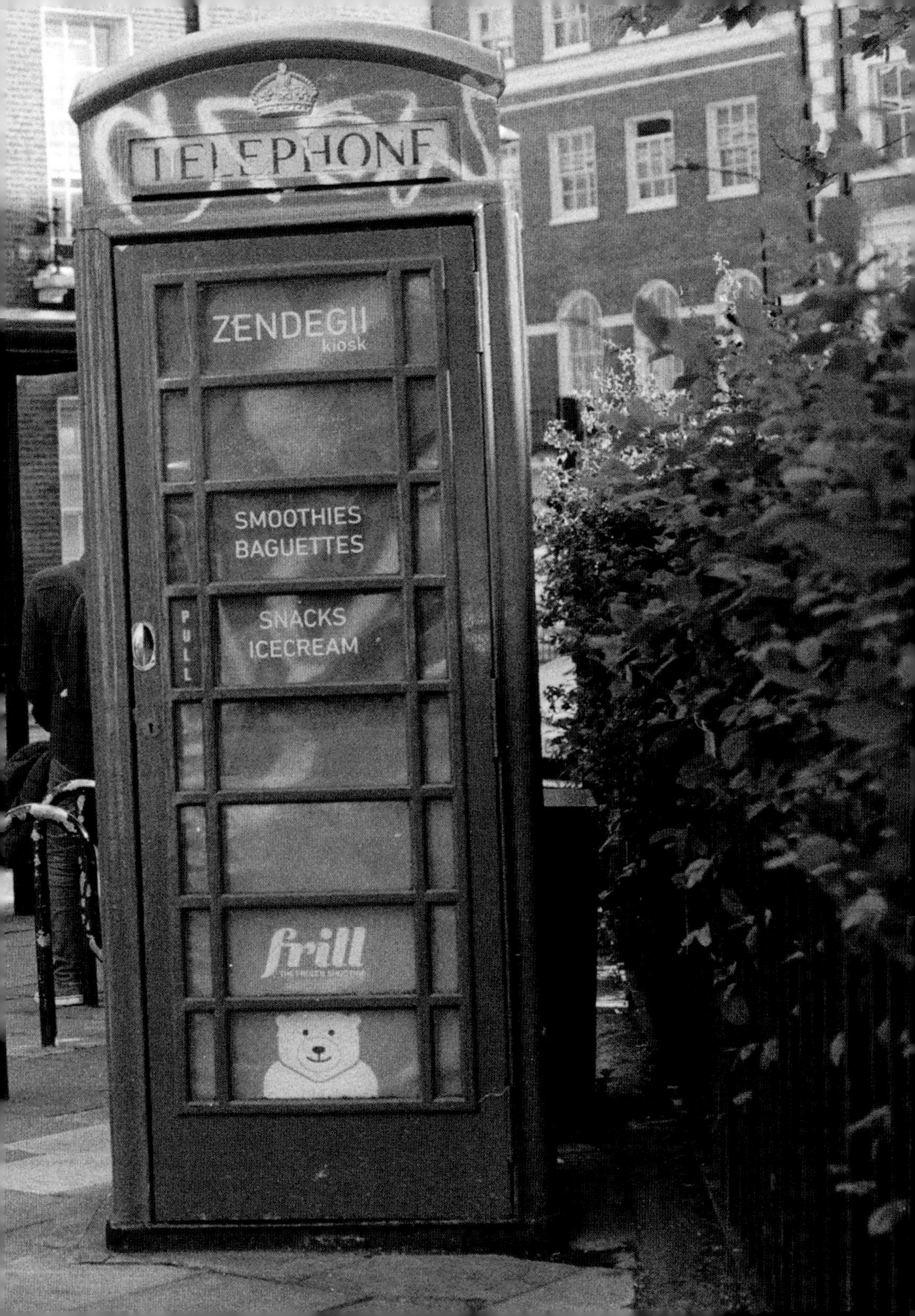
TELEPHONE
ZENDEGII
kiosk
SMOOTHIES
BAGUETTES
PULL
SNACKS
ICECREAM
frill

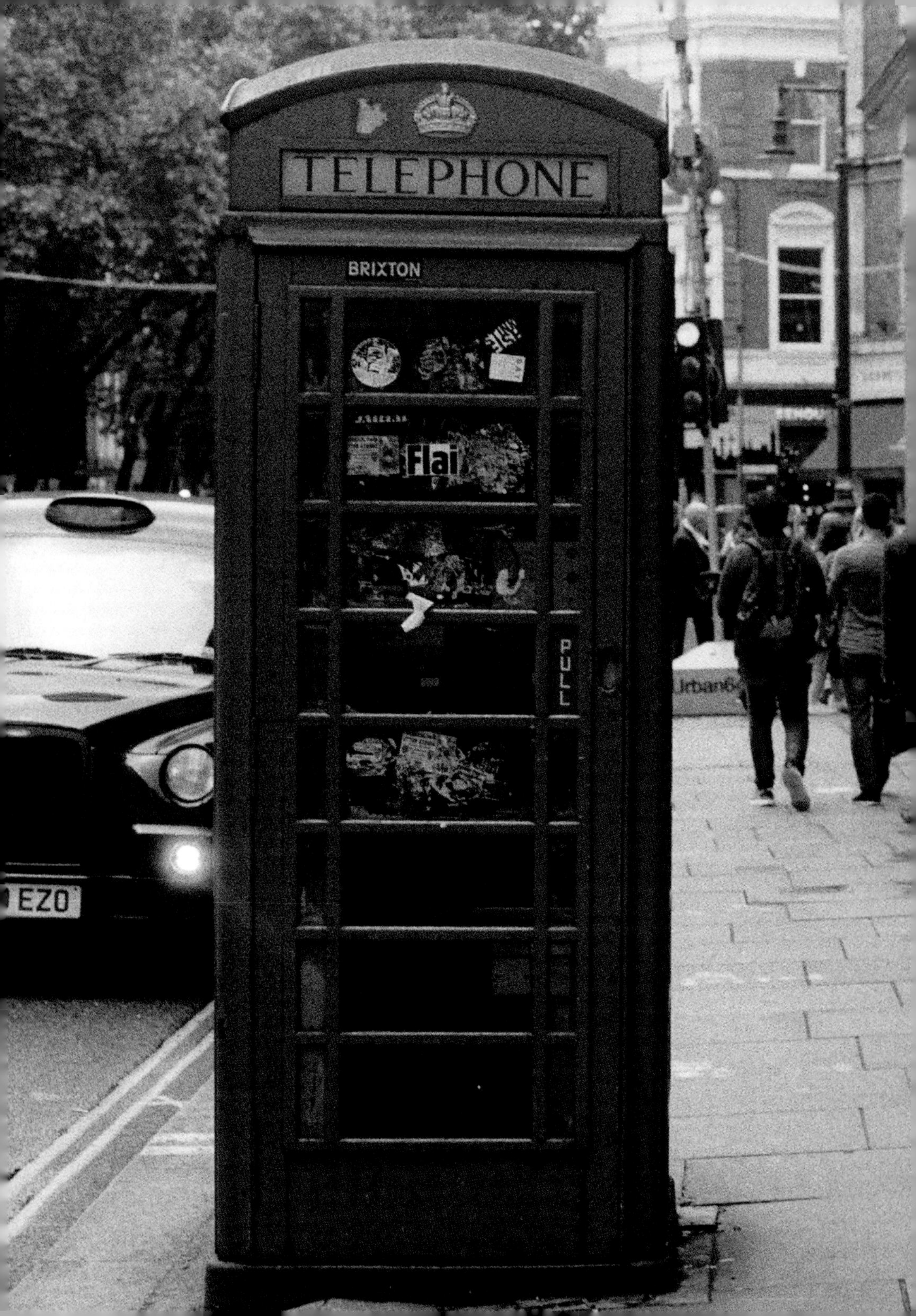

TELEPHONE
BRIXTON
Flai
PULL
EZO

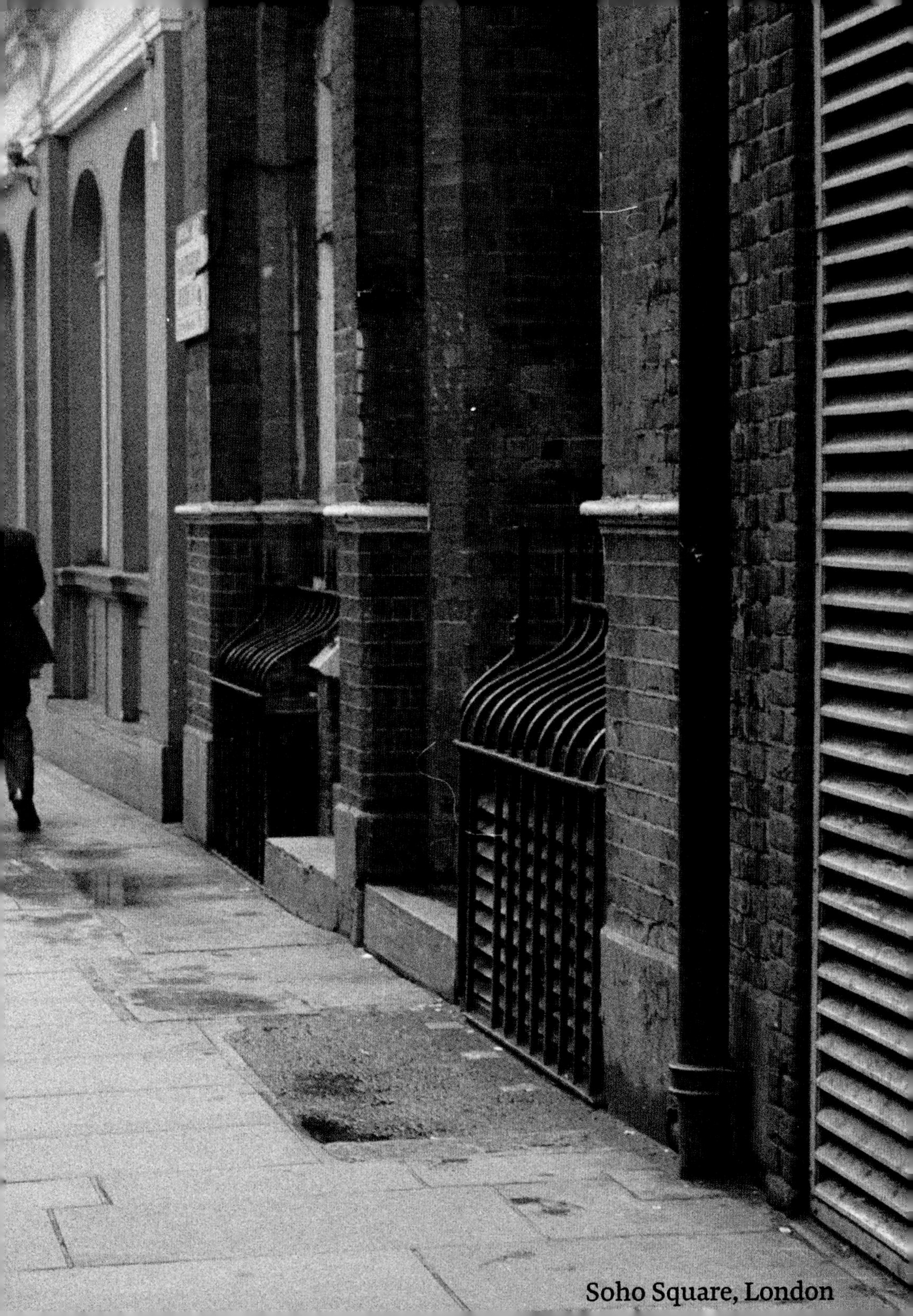

Soho Square, London

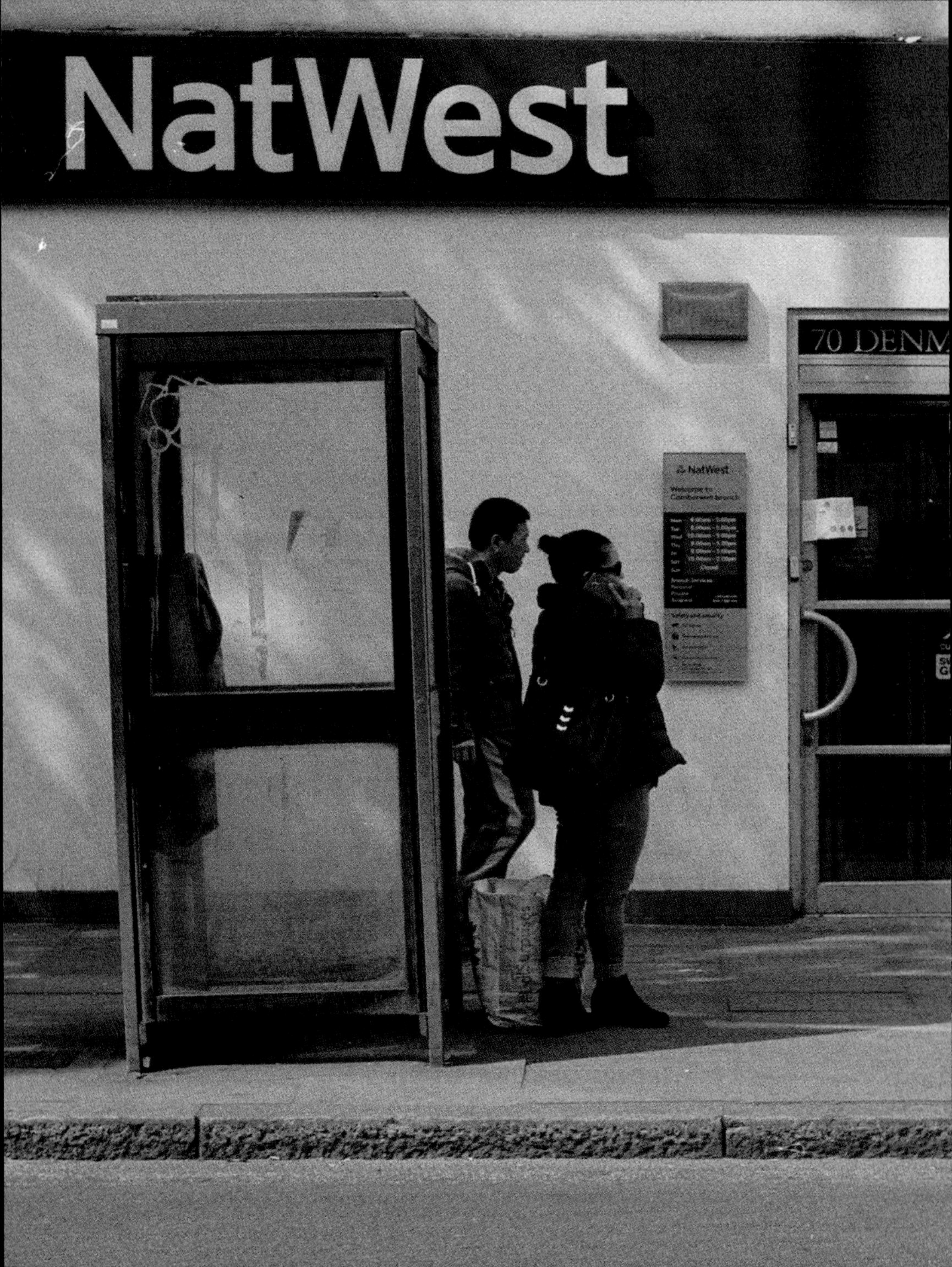

Denmark Hill, London

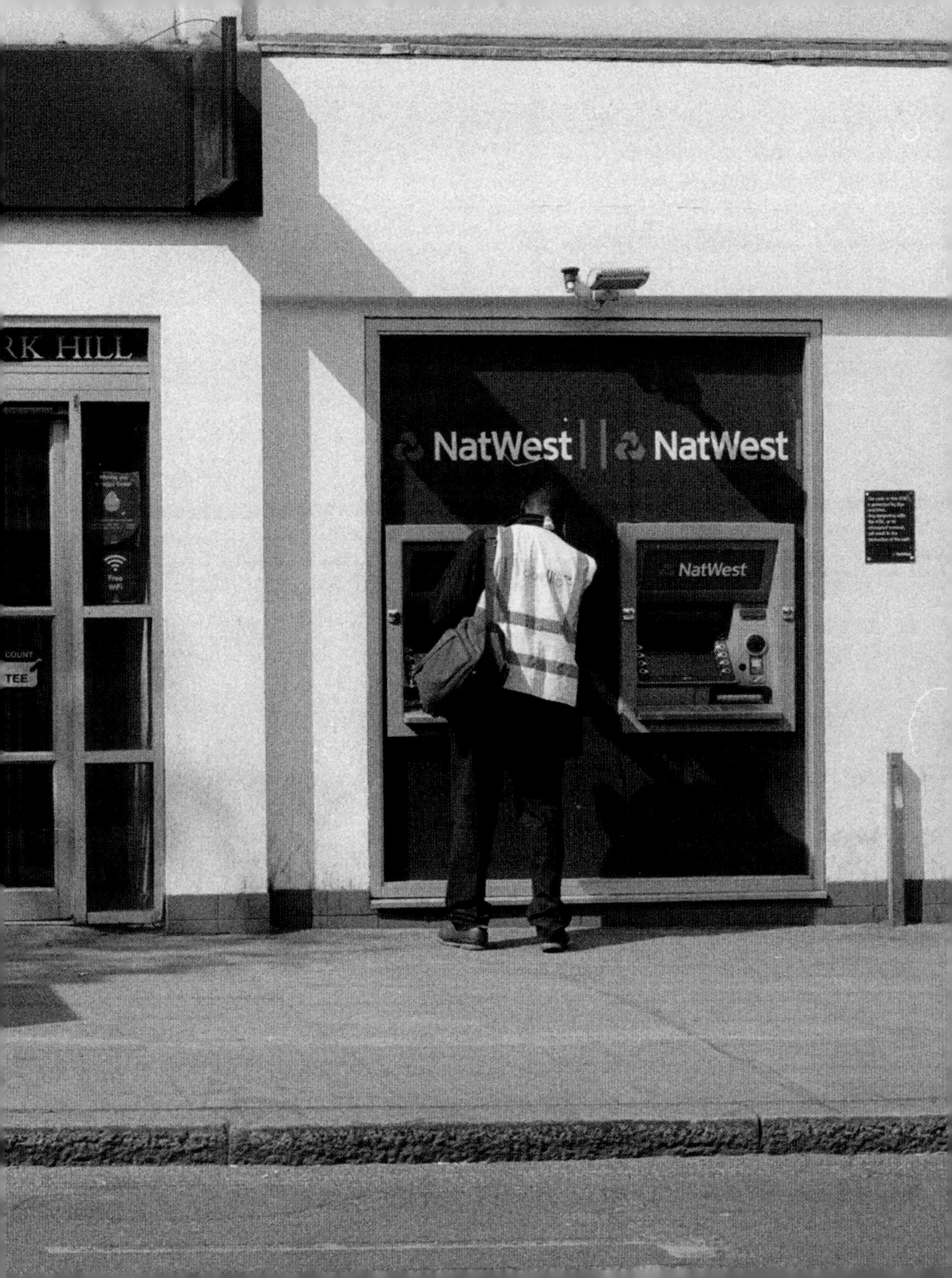
RK HILL
NatWest
NatWest
NatWest

Decommissioned

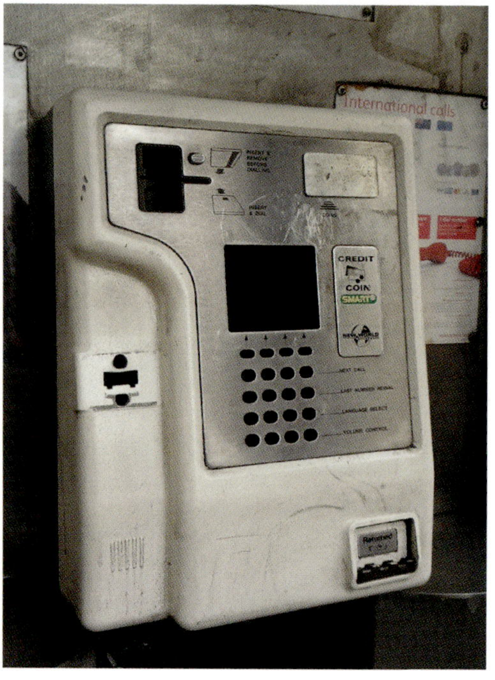
International calls
CREDIT
COIN

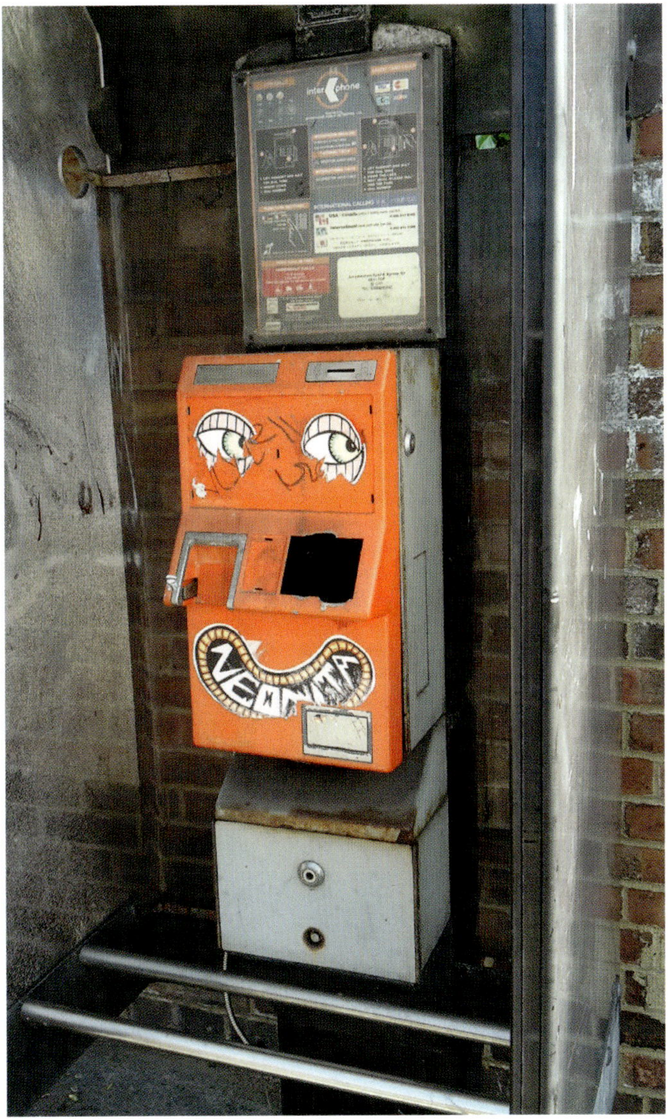
inter phone

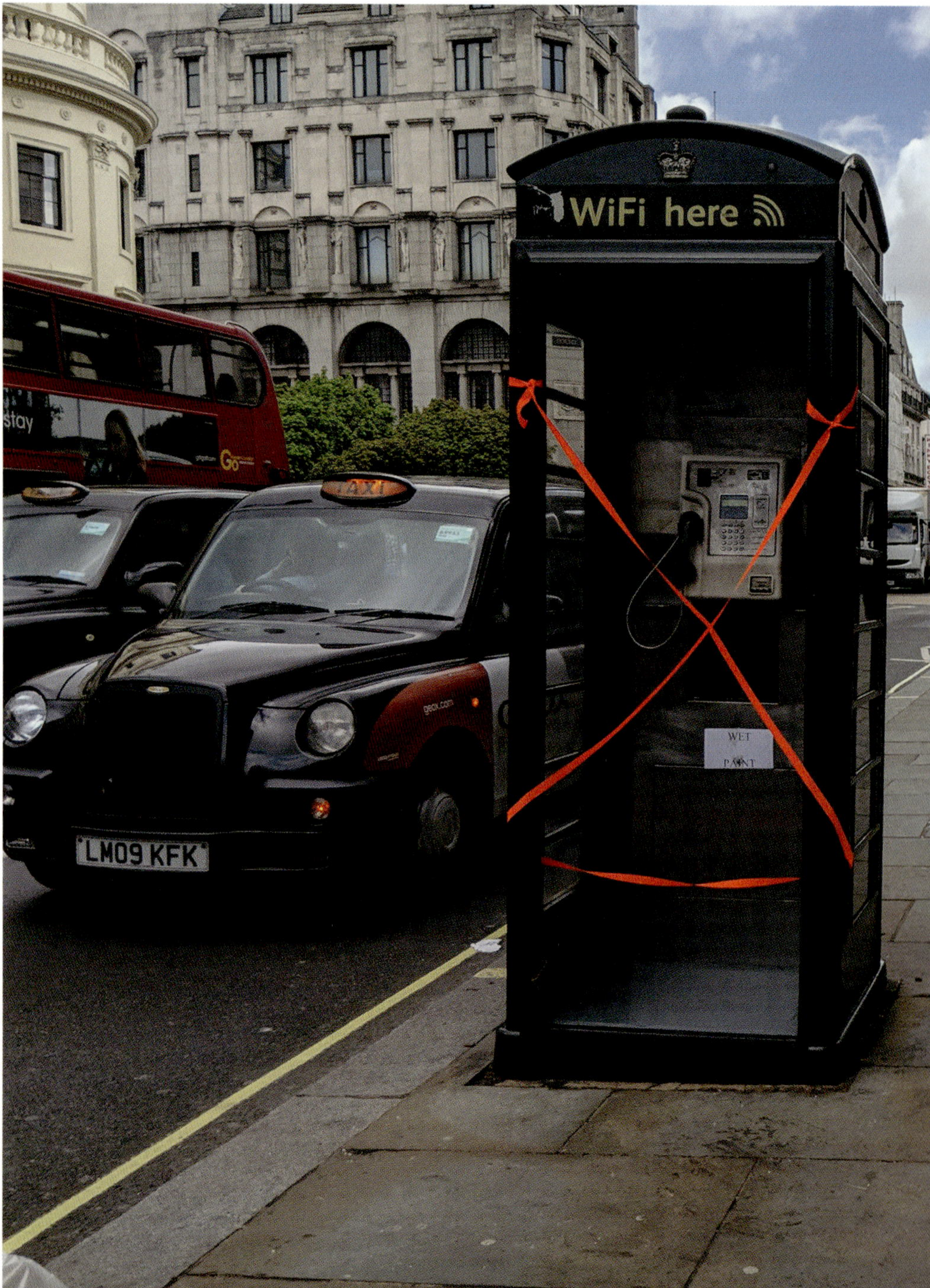

The Strand, London

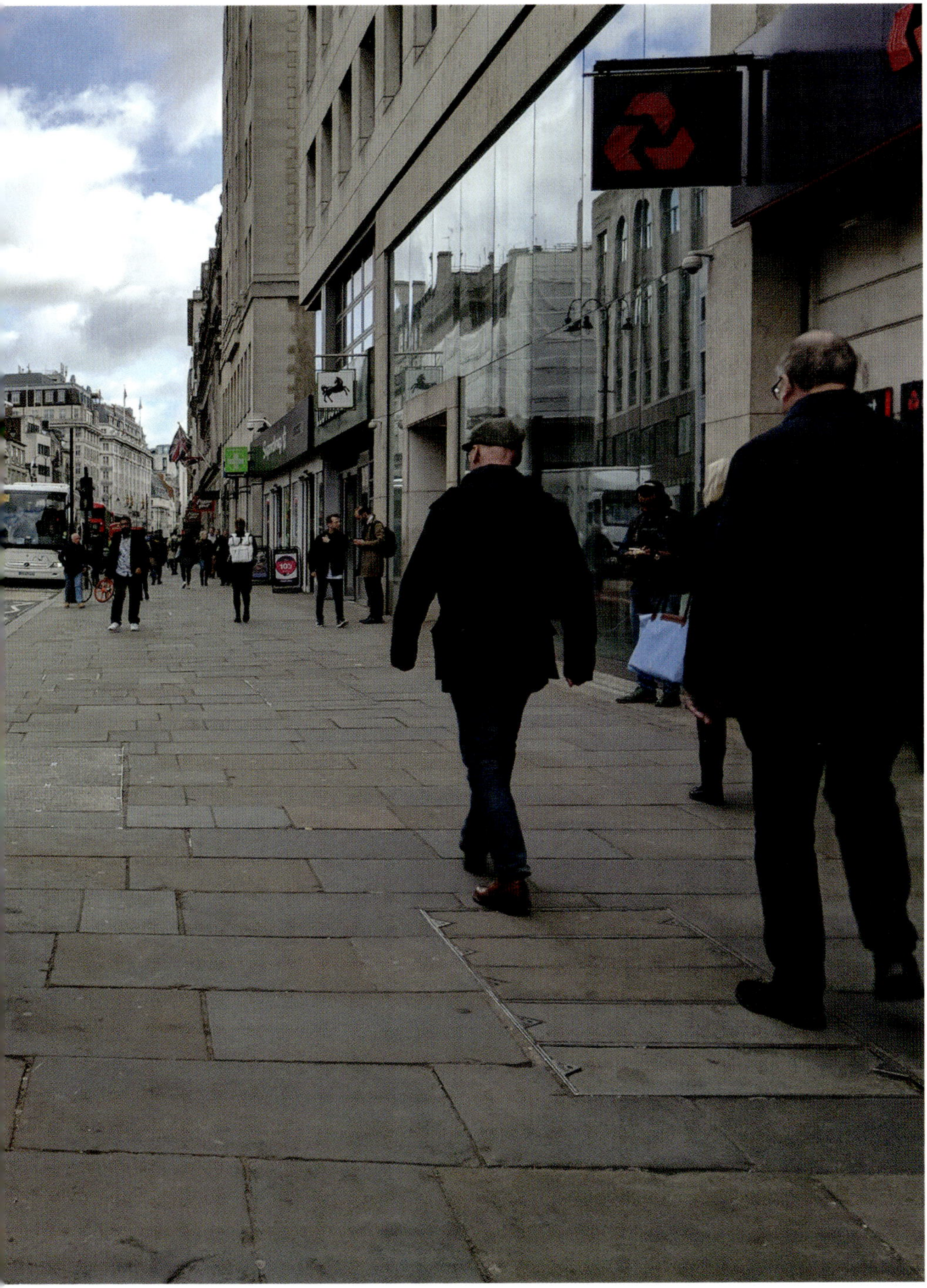

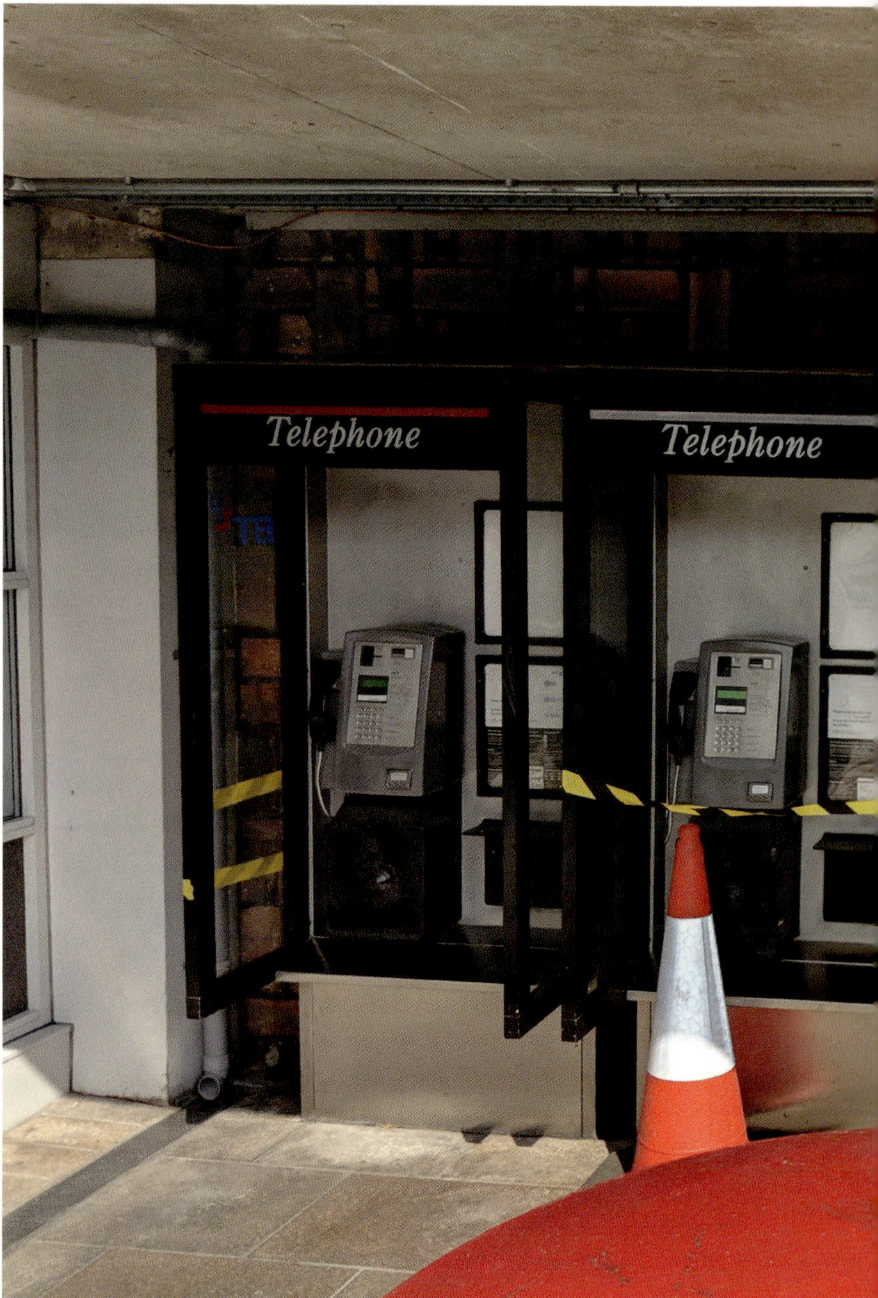
Telephone
Telephone

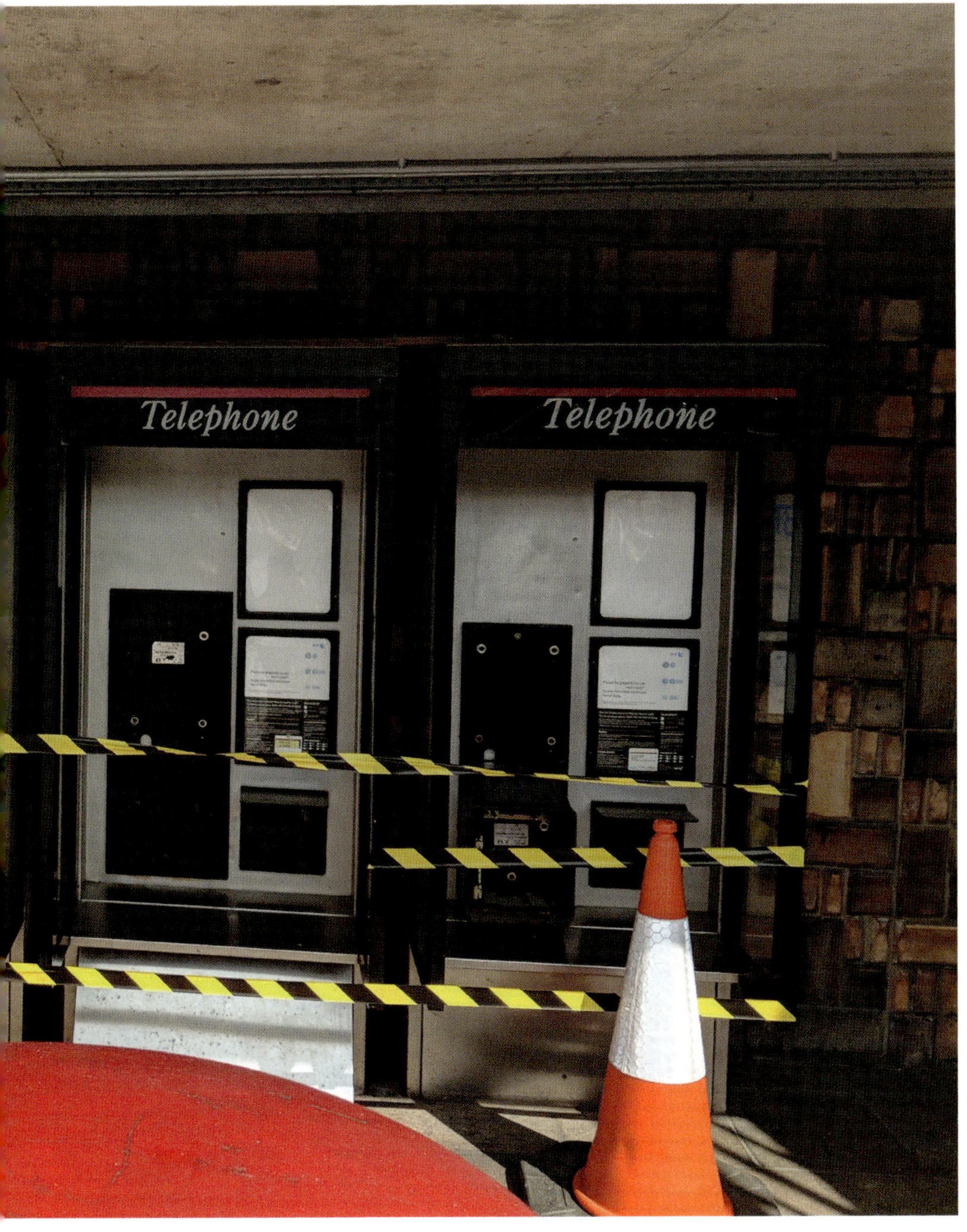
Telephone
Telephone

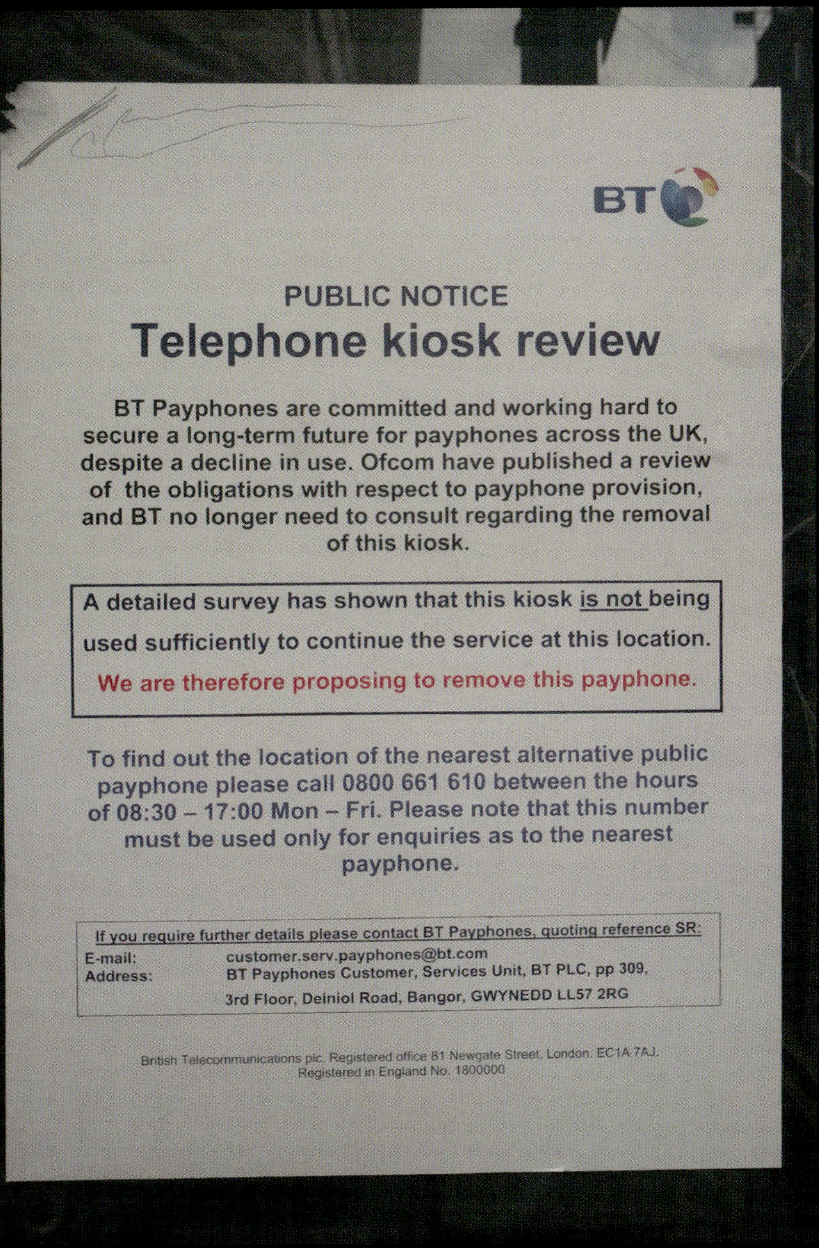

Near Oxford Street, London

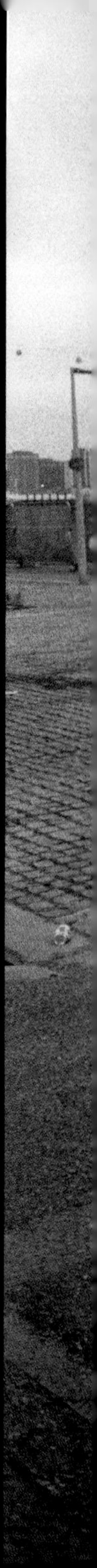

Greenwich Penninsula, London

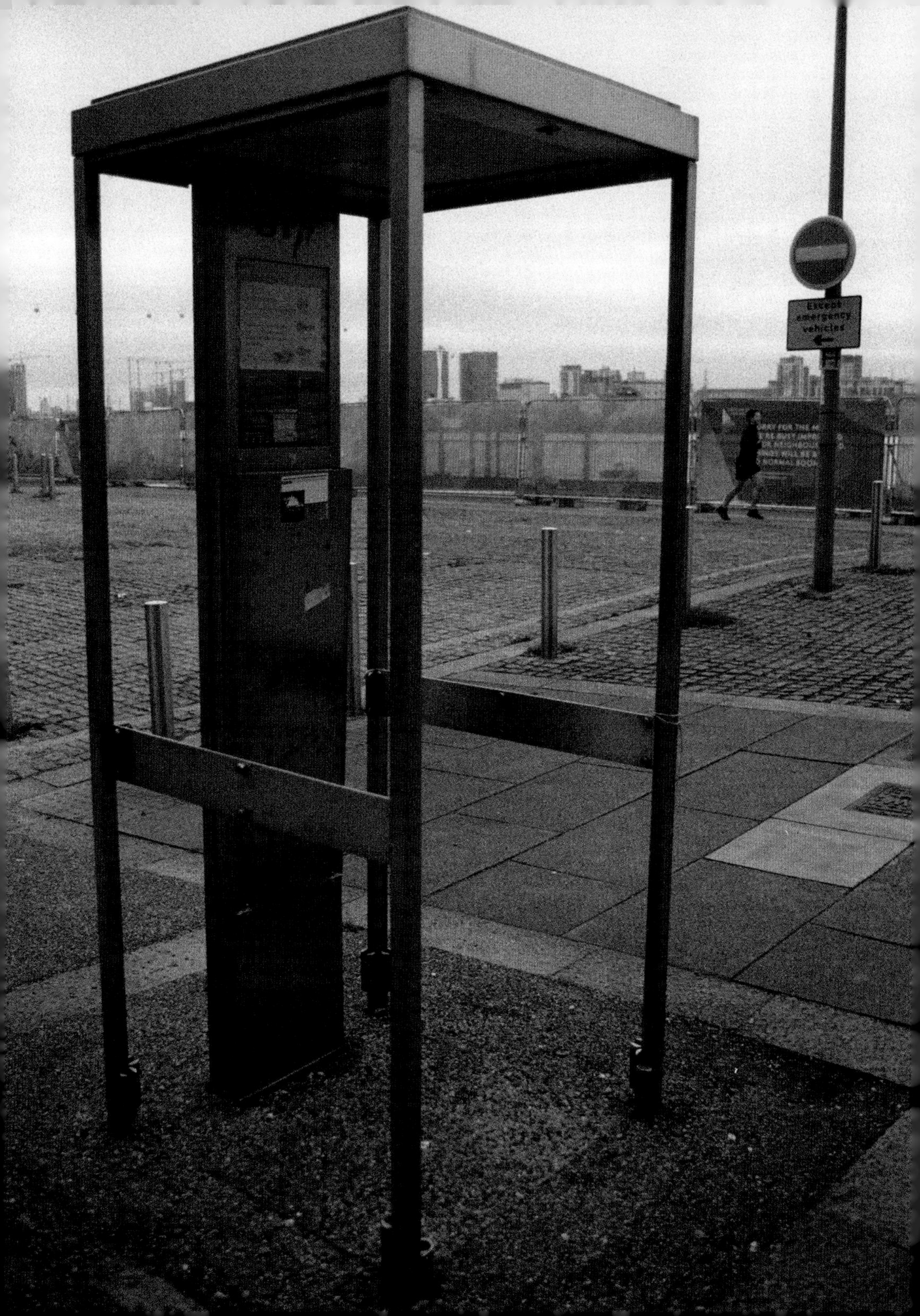
Except
emergency
vehicles

Future

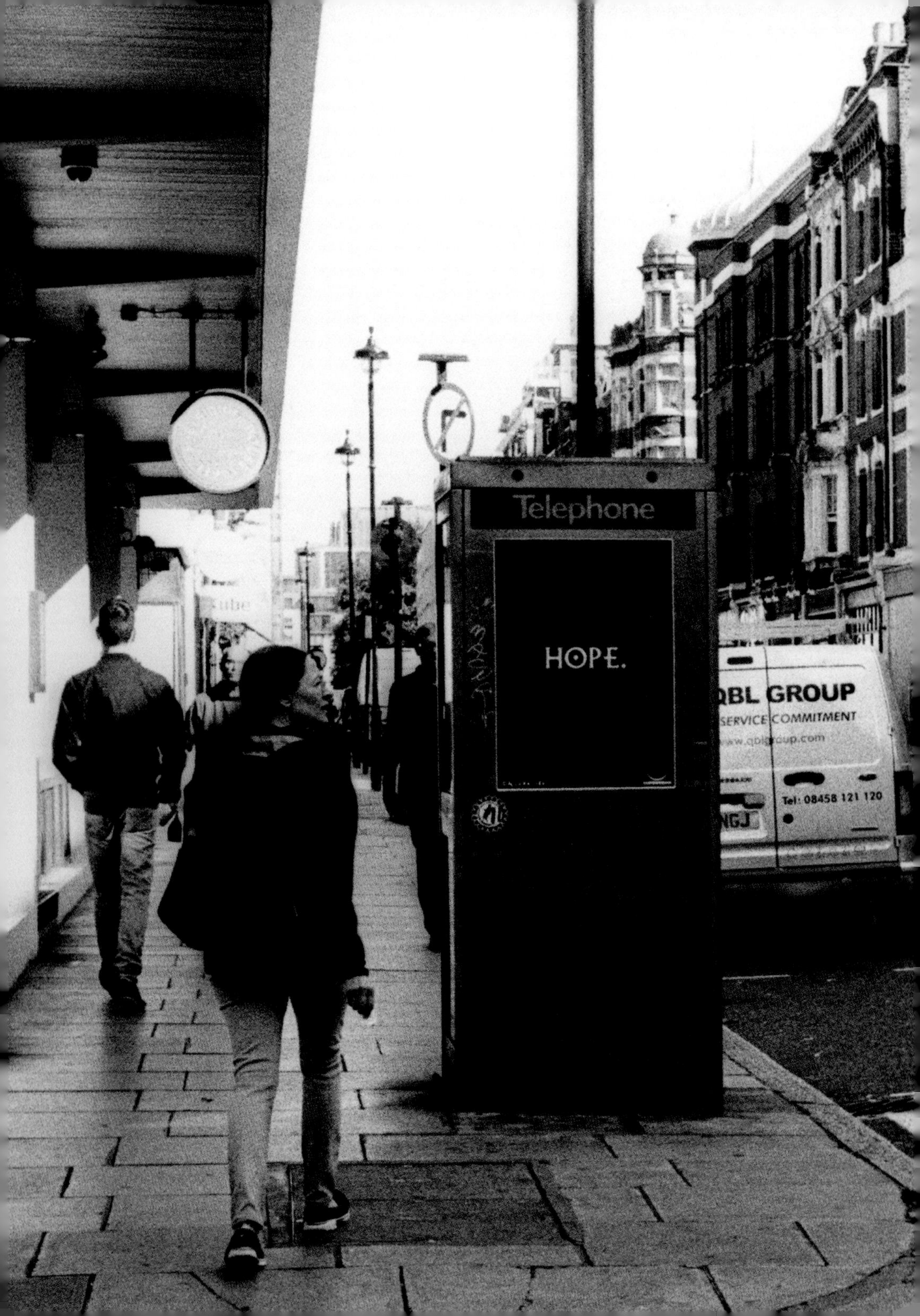
Telephone
HOPE.
QBL GROUP
SERVICE COMMITMENT
Tel: 08458 121 120

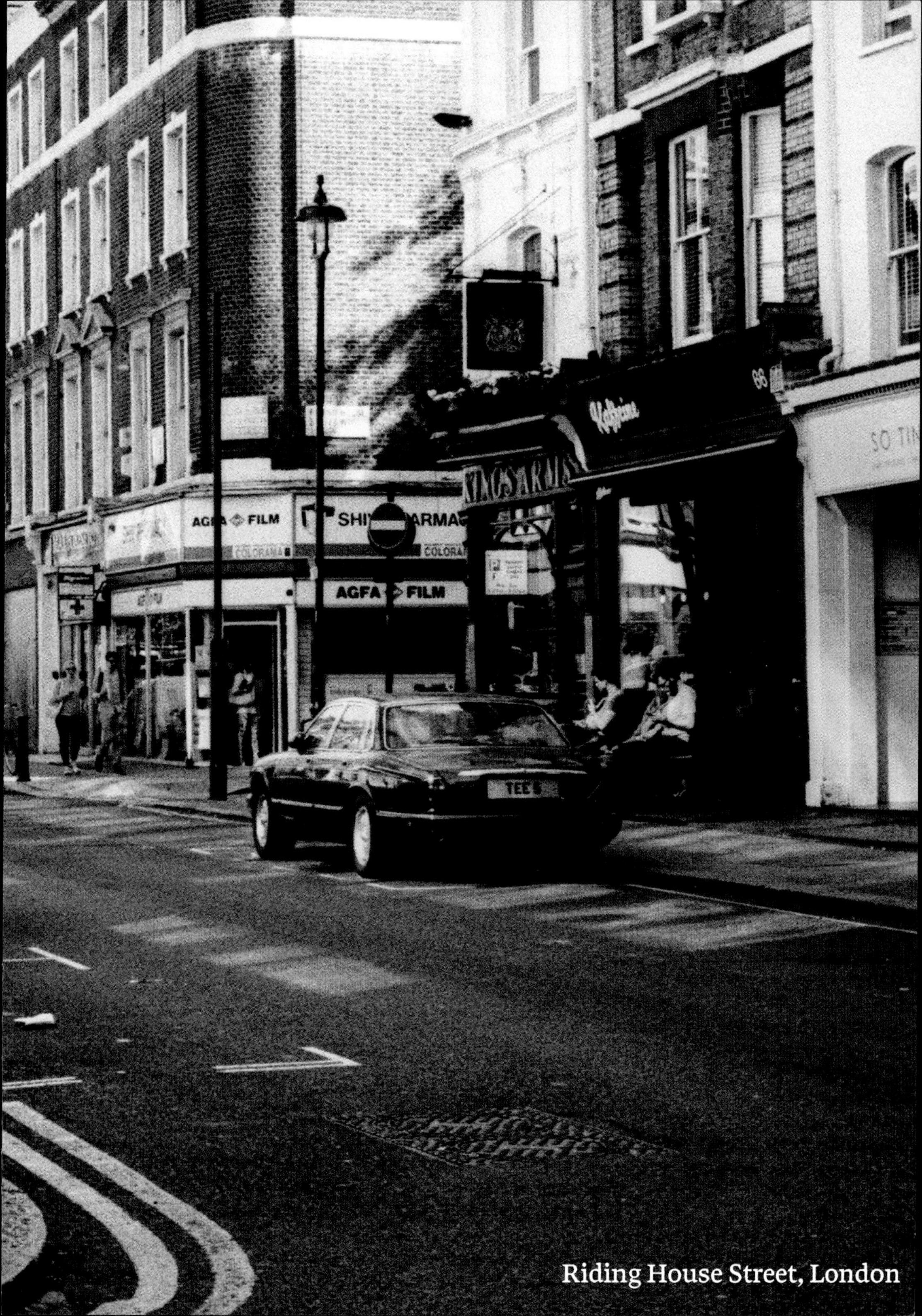

Riding House Street, London

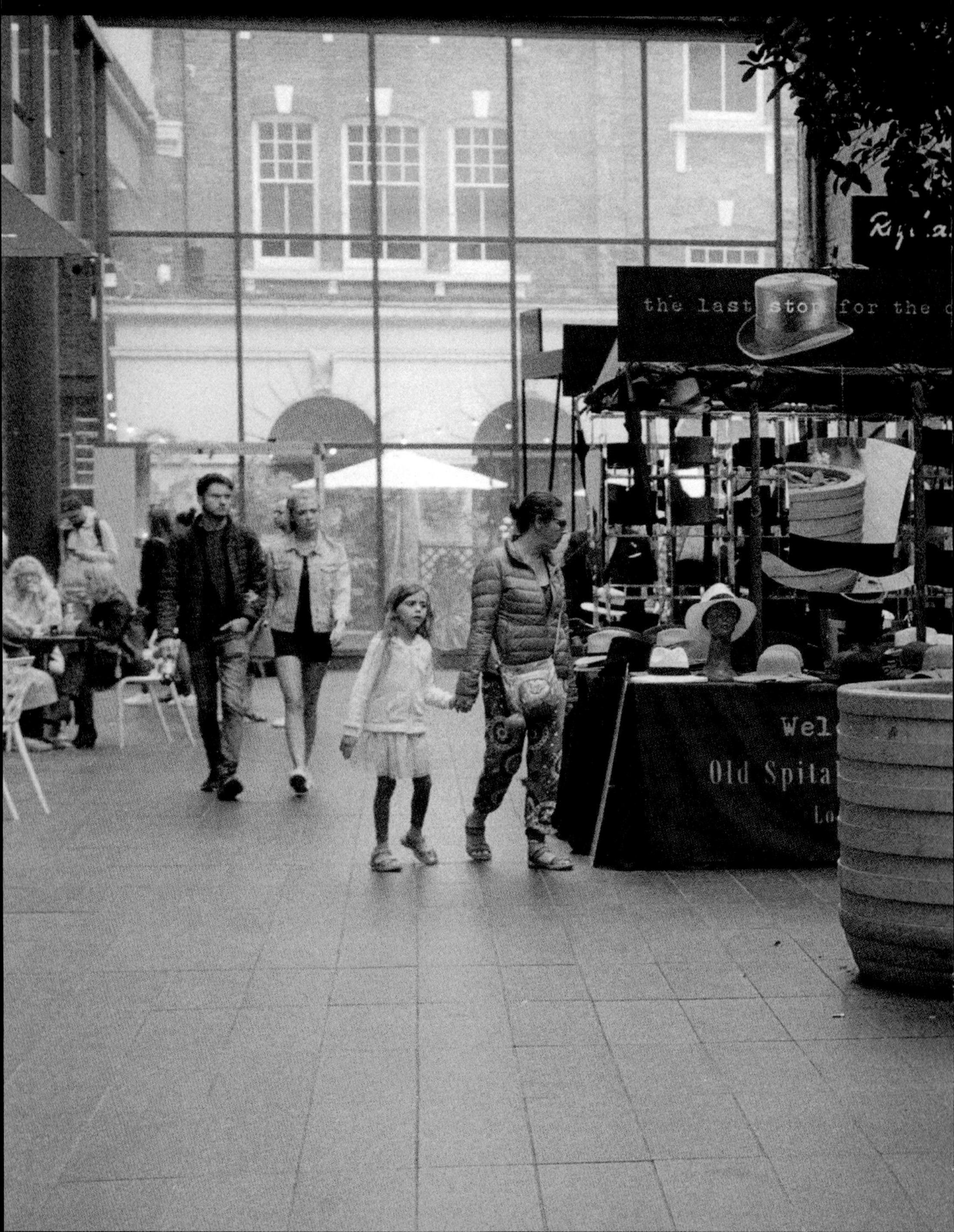
the last for the

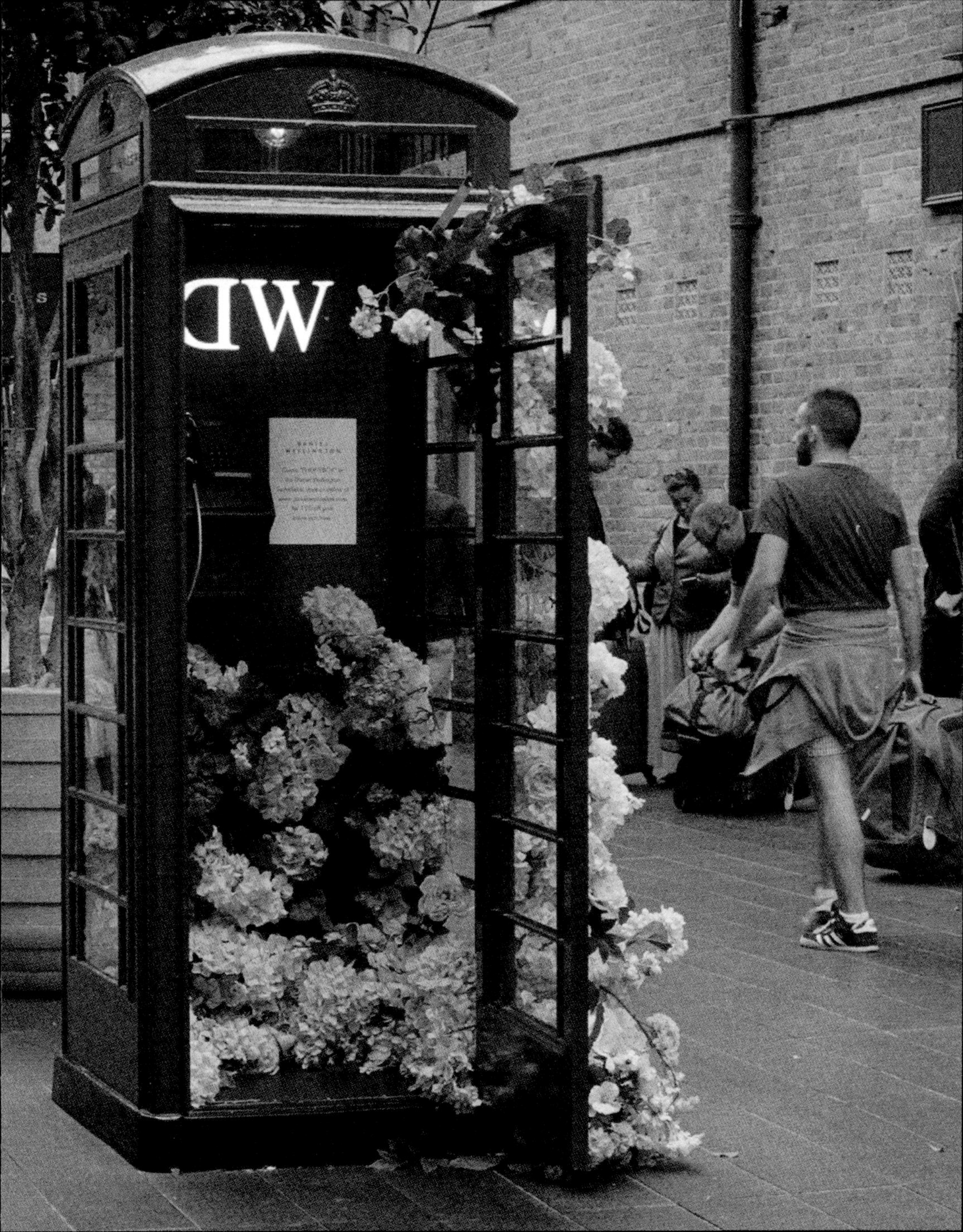

Spitalfields Market, London

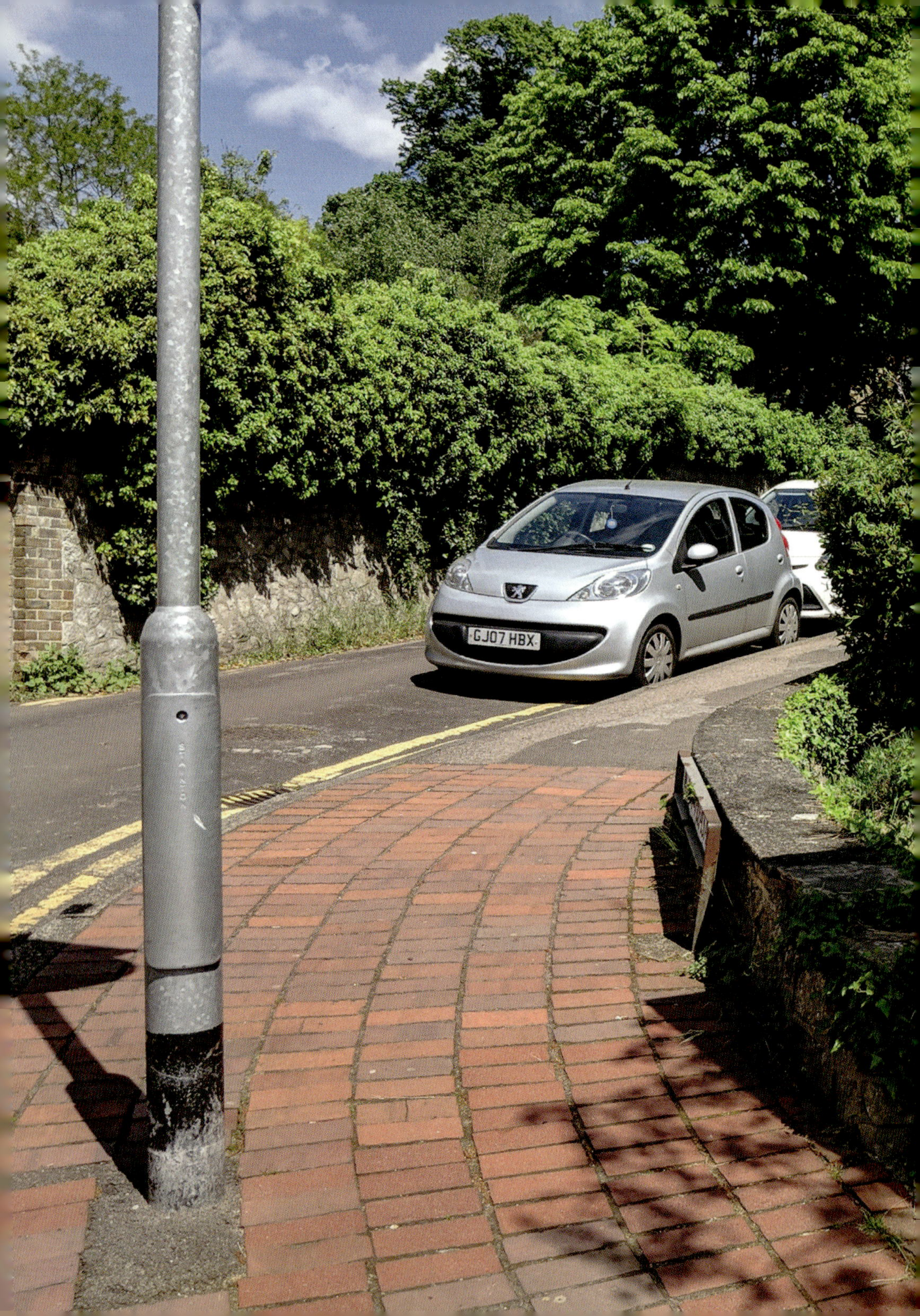
GJ07 HBX

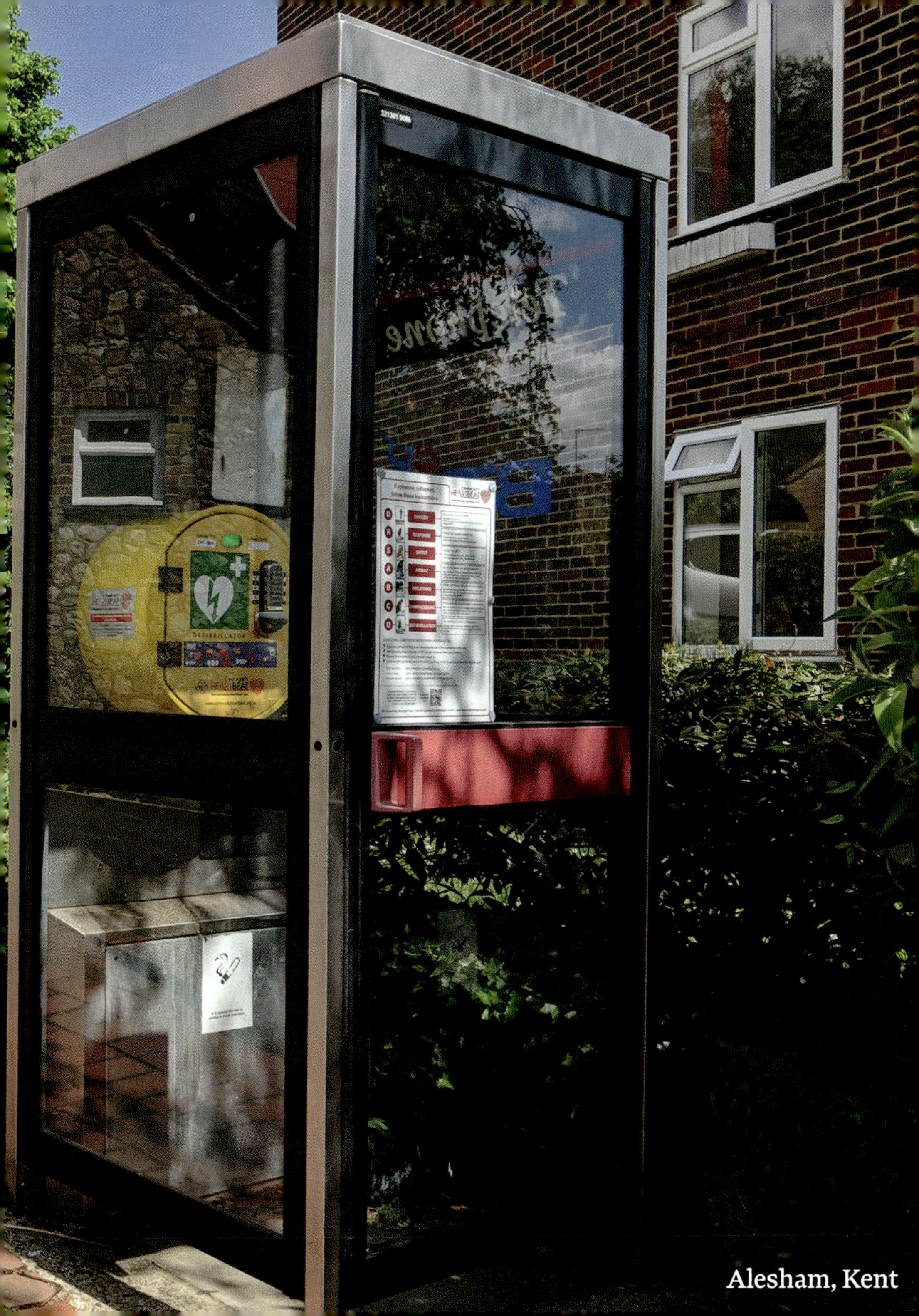

Alesham, Kent

"Any recognised local authority, parish/ community/town council, Registered Charity or any Community Interest Company throughout the UK or Scotland can apply to adopt their local phone box"

- BT Adopt a Kiosk Scheme

Whilst most kiosks will, eventually, be removed or replaced with objects which are not telephone kiosks - community groups have the opportunity to adopt them to improve their local area.

This scheme has been used to provide defibrillators (previous page) in remote locations where medical help may be a long way away (Community Heartbeat Trust), or to create tiny art galleries showcasing local artist's work or to build micro libraries (opposite) to serve local readers.

Others still will be acquired by businesses for profit, including coffee shops, , art installations, galleries, (the most micro of) micro pubs, phone repair stalls , and even, perhaps ironically, mobile phone charging stations.

It's clear, now, that these final remnants of communication history are now being handed over to society. It's up to us to decide what happens to them next.

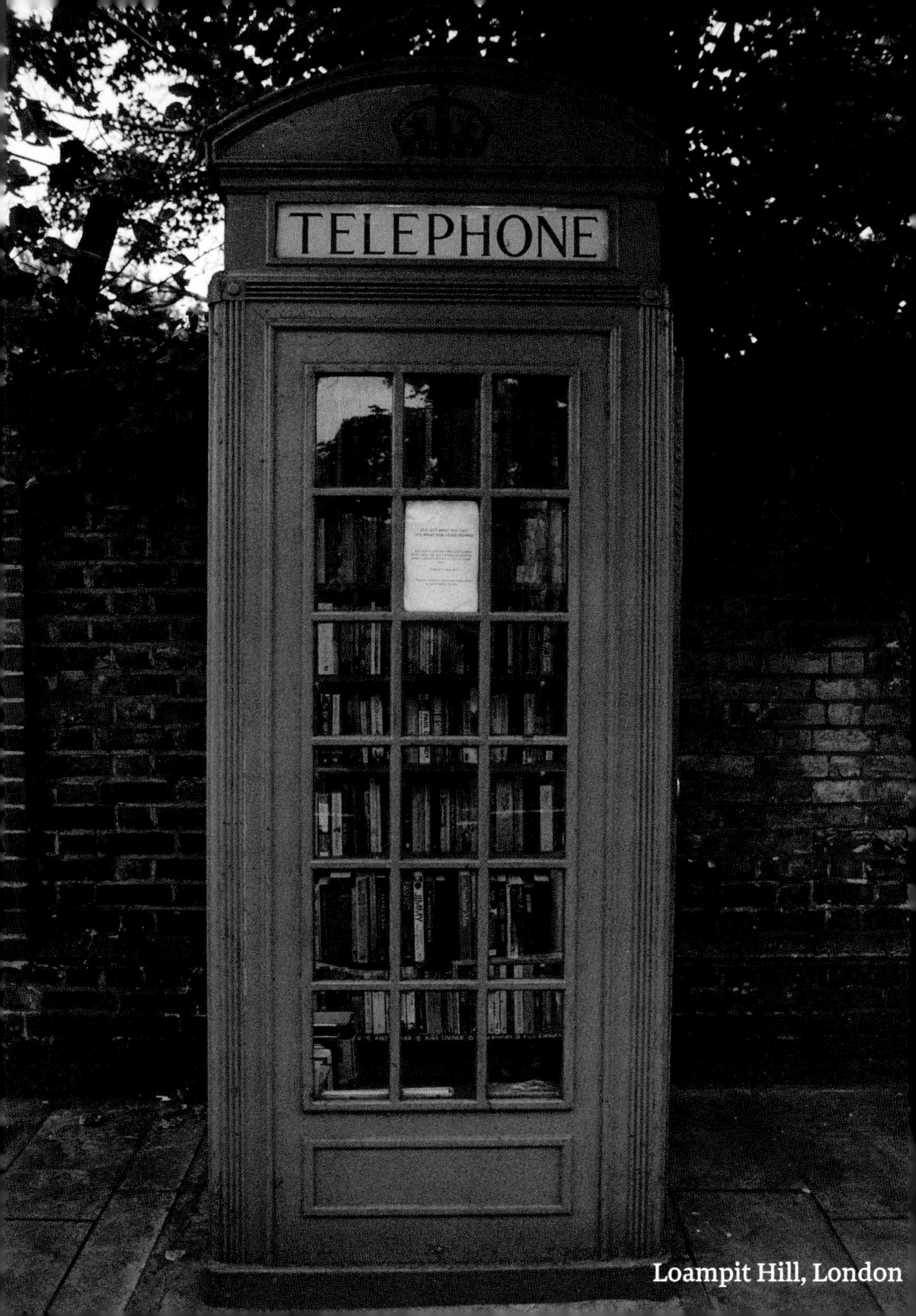

Loampit Hill, London

List of Images

Afterword

This project has been ongoing since 2012. At least, I think that is the point at which I started taking photographs of phone kiosks.

The original purpose of taking these photographs was to build a database of payphones and their respective numbers for Open Street Maps - many of which have indeed ended up in their dataset. However, after a while I realised that I was building up a catalogue of images of a piece of history that was soon to be lost forever. It is my hope that at some point - someone will find these 'zines' - and see what happened to the telephone kiosk towards the end of it's life.

Much in the world of technology and product development is focussed on bringing the service to market, launching it, and having it widely adopted. Little thought was traditionally made for what happens afterwards. These days sustainability is at the top of the agenda, and we must think about this more carefully. It is useful then to see what issues arise in that products that are now in this phase of their life cycle. Being "built to last" is not always a good thing.

These zines were produced during the COVID-19 pandemic of 2020 (at least, that's what I hope it's known as in the future...). Like many things, I expect the demise of the telephone kiosk is to be accelerated further by increased sanitary concerns with public infrastructure. So, this may be even more final a record of their existence than I had originally intended!

Thank you for reading!

6/40

Kiosk, Volume 2

Edition of 40 copies

First edition June 2020

Published by naxxfish photography

ISBN: 978-1-8380381-1-3

naxxfish.photography

naxxfish